Physical and Chemical Components of Wood

Their Influence on the Physical-Mechanical Properties of Ten Brazilian Wood Species

Bárbara Branquinho Duarte
Francisco Antonio Rocco Lahr
Antonio Aprígio da Silva Curvelo
André Luis Christoforo

ELIVA PRESS

ELIVA PRESS

Bárbara Branquinho Duarte
Francisco Antonio Rocco Lahr
Antonio Aprígio da Silva Curvelo
André Luis Christoforo

We have made full characterization of ten tropical wood species, including chemical composition. The statistical analysis by multivariate regression models of our results indicated some relationships, as the coefficient of determination (R^2) and analysis f variance ANOVA indicated that the equations for 10 of 16 properties are significant to estimate physical and mechanical properties of wood.

Published by Eliva Press SRL
Address: MD-2060, bd.Cuza-Voda, 1/4, of. 21 Chişinău, Republica
Moldova
Email: info@elivapress.com
Website: www.elivapress.com

ISBN: 978-1-63648-130-2

PHYSICAL AND CHEMICAL COMPONENTS OF WOOD –
THEIR INFLUENCE ON THE PHYSICAL-MECHANICAL PROPERTIES OF TEN BRAZILIAN WOOD SPECIES

B. B. Duarte*[a], F. A. Rocco Lahr[a], A. A. da S. Curvelo[b], A. L. Christoforo[c]

[a] Escola de Engenharia de São Carlos, Universidade de São Paulo, Departamento de Engenharia de Estruturas, Av. Trabalhador São-carlense, 400, CEP 13560-970, São Carlos (SP), Brasil.

[b] Universidade Federal de São Carlos, Departamento de Engenharia Civil, Rod. Washington Luís, km 235 - SP310, CEP 13565-905, São Carlos (SP), Brasil.

[c] Instituto de Química de São Carlos, Universidade de São Paulo, Av. Trabalhador São-carlense, 400, CEP 13560-970, São Carlos (SP), Brasil.

* Corresponding author at LaMEM - Laboratório de Madeiras e de Estruturas de Madeira - Universidade de São Paulo, Av. Trabalhador São-carlense, 400, CP: 780, 13560-970, São Carlos, SP, Brasil.

E-mail address: barbarabbduarte@gmail.com (B. B. Duarte).

DUARTE, Bárbara Branquinho; LAHR, Francisco Antonio Rocco; CURVELO, Antonio Aprigio da Silva and CHRISTOFORO, André Luis. Influence of Physical and Chemical Components on the Physical-Mechanical Properties of Ten Brazilian Wood Species. Mat. Res. [online]. 2020, vol.23, n.2 [cited 2021-02-16], e20190325.
Available from: <http://www.scielo.br/scielo.php?script=sci_arttext&pid=S1516-143920200002002 06&lng=en&nrm=iso>.
Epub May 25, 2020. ISSN 1980-5373. http://dx.doi.org/10.1590/1980-5373-mr-2019-0325.

Table of content

Abstract 3

1. Introduction 4

2. Materials and methods...7

3. Results 13

4. Discussion...26

5. Conclusions...29

6. Acknowledgments 29

References 30

Abstract

Wood offers a good combination of high strength and low density, but its complete characterization takes a long time. This study aimed to correlate apparent density, chemical component and porosity parameters with the physical and mechanical properties of the wood as an alternative route to reduce the time spent proposing equations to estimate each mechanical property. Ten tropical wood species were characterized in accordance with Brazilian Standards and chemical components were determined by the Klason method, and the porosity using mercury intrusion. Multivariate regression models were applied to the results of each species to find the relationships. Good results were obtained, such as the coefficient of determination and the analysis of variance, which indicated that the equations for 10 out of the 16 properties are significant. Therefore, the apparent density, chemical component and porosity parameters used reduce the time intervals much lower than the time test stipulated by the standard.

Keywords: Physical and mechanical properties; density; porosity; cellulose; hemicellulose; lignin; ash.

1. Introduction

Wood is widely used in several sectors, mainly in civil construction, because wood combines high strength, low density and low cost. Consequently, it is widely used for structural applications [1, 2, 3].

In North America (USA and Canada), wood is found in most residential houses and commercial buildings. Even though they are built out of brick or stone walls, they have elements with wooden structures. As a result, housing construction consumes (about) a third of the wood-based products in the USA [4, 5]. In the European market, wood has been more widely used recently in construction, and other areas, due to increasing availability of knowledge, as well as public policies that support the use of wood products, resulting in lower energy consumption and reduced CO_2 emissions compared to other materials (concrete, brick or steel) [6].

In Brazil, wood is one of the most traditional materials used in construction. It is used for building components, such as roofing structures, floors, window frames, and in walkways and bridges. According to the National Forest Information System (NFIS), in 2014 the total production of sawn lumber in Brazil was approximately 11 million m³ [7, 8].

Wood consumption and wood products have grown gradually in recent years in Brazil, mainly due to financing from governmental institutions (SiNAT Directive N005 / 2011) [9, 10]. The Brazilian standard - NBR 7190 [11] is the main guiding principle for projects and studies on wood, which establishes the general conditions that must be followed in the design, execution, and control of current wooden structures, such as bridges, pontoons, roofs, floors and structures. This standard presents six properties required for dimensioning structural elements for 43 species of hardwood (dicotyledonous angiosperms) and softwood (gymnosperms or conifers).

It is estimated that in the Amazon Forest, there are approximately 12,655 species, many of which are being introduced into the market, and only about

7,694 are properly cataloged and characterized, which greatly restricts their potential for applications. Thus, it is very important to characterize new species [12].

Wood extracted in Brazil can be used in several sectors, but often it has been sold under generic names and, thus, without any characterization, which can negatively influence its performance in service (ALMEIDA, 2015). [13].

Although it is important to know the physical and mechanical properties of wood, they are obtained from running a long series of tests and using various kinds of equipment, available only in large research centers, which makes it difficult to determine them. According to Ter Steege et al [12], it would take at least 300 years to catalog and characterize all the tree species in the Amazon [14].

For each species to be studied, 14 mechanical tests are performed, described in the Brazilian standard [11], which requires 12 specimens for each test, totaling 168 specimens. In addition, the time required to perform the tests is approximately three weeks for each species. It should also be taken into account that in order to prepare the specimens, there is a need to remove a batch containing several logs and send it to the research center where tests will be carried out. On the other hand, in order to know the chemical composition of a wood species, analysis requires approximately 15g of ground or particulate material from the wood. The estimated time is three days to carry out the analysis. Mechanical tests need a larger amount of material to be removed, making transportation more difficult and costly, in addition to needing more time for its performance.

The Brazilian standard [11] shows relationships for estimating strength and stiffness from other wood properties, however there are no equations that associate the physical and mechanical properties of wood with its chemical components [12, 15]. For example, Dias et al. [16] considered the results obtained for forty tropical wood species in Brazil and obtained equations from which dependence among apparent density (12% moisture content) and strength and stiffness properties can be observed. However, the results of the regression models in property estimation did not provide good adjustments as the values for the coefficient of determination (R^2) were, in most cases, below 0.80. The authors recommend inserting new parameters to improve the mechanical property estimates.

Fujimoto et. al. [17] evaluated the possibility of estimating mechanical properties using Near Infrared (NIR) spectroscopy. From samples of wood native to northern Japan (*Larix gmelinii* var. *japonica* and *Larix kaempferi*), the authors carried out mechanical and physical tests to determine the modulus of elasticity, modulus of rupture, compression strength parallel to the grain, dynamic modulus of elasticity of the air-dried wood and density following the Japanese standard JIS Z 2101.

As previously mentioned, wood extracted in Brazil offers possibilities to be used in various different sectors, however it was often commercialized without any characterization, which shows there is a great need for this study. Thus, this study aimed to use Brazilian wood species that have not yet been fully characterized and are already being commercialized.

Considering the limitations inherent to the studies reported in the literature to date, this study aimed to correlate apparent density, chemical component and porosity parameters with physical and mechanical properties of wood, using multivariable regression models.

2. Materials and methods

Table 1 shows the wood species studied with the respective images of the samples. The wood was collected from a company called *Madeireira César Ind. e Com. Madeiras Nobres*, located in Brotas, São Paulo, Brazil. The specimens were cut, removing them from regions that were at least 30 cm away from the ends of the piece on each side.

Table 1. Ten wood species and its illustrative image.

Common Name	Scientific name	Sample Image
Tatajuba	*Bagasse guianensis*	
Roxinho	*Peltogyne recifencis*	
Cambará	*Erisma uncinatum*	
Cedroarana	*Cedrelinga catenaeformis*	
Cumaru	*Dipteryx odorata*	
Cupiúba	*Goupia glaba*	
Caixeta	*Simarouba versicolor*	
Cedro	*Cedrela fissilis*	
Sucupira	*Diplotropis purpúrea*	
Cajueiro	*Anacardium* sp.	

It should be mentioned that 12 pieces of each species were needed for the tests. For this purpose, the company used 6 trees from each species to remove the pieces, two from Macajaí, Roraima (RR), two from Cláudia, Mato Grosso (MT) and two from Maués, Amazonas (AM). Each tree was first cut above the diameter at breast height, approximately 1.3m from the base, and the second cut was made 3m from the first, therefore 2 pieces were removed from each tree, totaling 12 pieces. Sampling was performed from wood obtained from axial and radial cuts.

In order to ensure that the study was fully representative, species were chosen that are grouped into the five resistance classes indicated by the

7

Brazilian standard [11] for dicotyledons (two species for each class), that is, it covers a wide range of densities. Thus, there are two species that would be within each of the five classes of dicotyledons (C20, C30, C40, C50 and C60).

According to Silva et al [18], the wood strength and stiffness are influenced by moisture content under the fiber saturation point, and consequently the specimens were submitted to an air conditioning room to reach the equilibrium moisture content, 12%.

The physical and mechanical properties of the samples were evaluated using the Brazilian standard [11], respectively. Table 2 lists the name of 17 parameters and their initials, respectively, to make it easier to distinguish between the physical and mechanical properties of the wood. Twelve wood specimens were used for the study to obtain the physical and mechanical properties, resulting in 2040 samples for the experimental tests.

Table 2. Physical and mechanical properties evaluated.

Property	Abbreviations
Apparent density	ρ_{ap}
Total radial retraction	TRR
Total tangential retraction	TTR
Compressive strength parallel to fibers	f_{c0}
Resistance to normal fiber compression	f_{c90}
Tensile strength parallel to fibers	f_{t0}
Resistance to normal fiber traction	f_{t90}
Shear strength parallel to fibers	f_{v0}
Cracking resistance	f_{s0}
Conventional resistance in the static bending test	f_M
Modulus of elasticity in compression parallel to fibers	E_{c0}
Modulus of elasticity in normal fiber compression	E_{c90}
Modulus of longitudinal elasticity in traction parallel to fibers	E_{t0}
Conventional modulus of elasticity in the static bending test	E_M
Hardness parallel to fibers	f_{H0}
Normal fiber hardness	f_{H90}
Tenacity	W

The chemical compositions evaluated consisted of cellulose, hemicellulose, lignin, extractives and ash contents. The sample preparation process followed the precepts of the TAPPI 264 CM-97 standard (1997) and the entire chemical analysis was performed in duplicate.

First, a Willey SL-31 knife mill was used to grind the material that was sieved using a 30-mesh size, thus obtaining 15 grams of sample per species of wood evaluated. The samples were made using filter paper containing half the material of each sample (7.5g) to perform the duplicate, which were subjected to the extraction process using organo-soluble solvents in a Soxhlet extractor. There were two steps. The first one consisted of using 1:1 (v/v) parts of cyclohexane/ethanol, for 8 hours to remove the organo-soluble extractives; and the second used boiling water for the same period. The materials were dried at 100 °C for 30 min in a kiln.

To determine the Lignin content of the material, the Klason method modified by NREL (2012) was used. This method was divided into two steps. The first step consisted of preparing the

an initial dry mass of 0.8 g of ground material (42 mesh) with 12 ml of 72% sulfuric acid added to an autoclavable borosilicate tube of 500 ml, which reacted under constant agitation for a period of 2 hours. The second step consists of adding 450 ml of distilled water to the tube (3% sulfuric acid concentration), which was then closed and placed in the autoclave at a temperature of 120 °C at a pressure bar of 2.0.

After the autoclave was depressurized, the tubes were cooled using ice and water to room temperature (25 °C). From this method, a residual solid and a solution were obtained, which were then filtered through a sintered glass funnel of known mass, using the retained material to calculate the insoluble lignin and filtered liquid to determine the soluble lignin. The sum of both the insoluble lignin and soluble lignin consists of the total lignin value of the material.

The insoluble lignin was obtained from samples of the retained material from the filtration. The retained material was oven dried for 24 hours and transferred to a desiccator to produce a constant mass. This dry mass is the insoluble lignin content. After that, the insoluble lignin content is correct for inorganic content. It should be mentioned that the presence of insoluble lignin ash should be considered, determined from the calcination of the residue obtained from the Klason method. Calcination consisted of exposing the retained material to 525 ° C for 4 hours with a heating ramp of 100 ° C.h⁻¹. The insoluble lignin (IL-%) was calculated by the following equation (1):

$$IL = \frac{m_L - m_C}{m_S} \tag{1}$$

where m_L is the lignin dry mass (g), m_C is the ash mass (g) and m_S is the dry sample mass (g).

The soluble lignin (SL) was obtained from liquid samples (7.5 ml) that were filtered and analyzed by Ultraviolet - UV-Vis Spectrophotometer (DR-5000) using a quartz cell of 10mm optical path, due to the lignin strongly absorbing the ultraviolet light characteristic by the aromatic characteristic. Spectrophotometry measurements were carried out in the region of 280 nm and 215 nm wavelengths, corresponding to ultraviolet light. A dilution factor of 5 was used, that is, 1 ml of the filtrate was diluted 4 times in the sulfuric acid solution in the same proportions. The soluble lignin is calculated by the following equation (2):

$$SL = \frac{(4.53 \cdot A_{215} - A_{280}) \cdot v_f \cdot DF}{3 \cdot m_b} \tag{2}$$

where A_{215} is the absorbance at 215nm, A_{280} is the absorbance at 280nm, vf is the measured volume of the acid solution (Klason reaction), DF is the dilution factor (dilution of the solution in which the spectrum was recorded) and m_b is the biomass dry mass used for the Klason reaction.

The cellulose and hemicelluloses contents were obtained by high performance liquid chromatography from the quantification of the sugars (and derivatives) contained in the hydrolyzate resulting from the determination of Lignin Klason. The quantified sugars and derivatives were: cellobiose, glucose, xylose, arabinose, acetic acid, furfural and hydroxymethylfurfural. In the chromatographic analyses (SHIMADZU–CR 7A), the chromatographic analysis of furfural and hydroxymethylfurfural is performed first.

The solution containing the hydrolysate obtained (Klason method) was filtered using a 0.45 µm membrane. The resulting solution was injected into a) a Hewlett-Packard RP18 (C18) column with a 1:8 (v:v) acetonitrile/water solution eluent, containing 1% acetic acid and a flow set at 0.8 mL min^{-1} at room temperature, using a UV– 254nm detector (SPD-10A-model). Then, the hydrolysate solution was filtered in membrane (SEP PAK C18 - Waters) to retain the soluble lignin and other degradation products. The second part is the chromatographic analysis for sugar determination, that the same solution containing the hydrolysate was filtered with a membrane of 0.45µm. The resulting solution was injected into the chromatograph through an Aminex HPX-87H column (300 × 7.8mm BIO-RAD) with an eluent 0.005 M H_2SO_4 and flow set at 0.6 mL/min and temperature at 45 °C, using a refractive index detector (SHIMADZU R10-6A).

In accordance with Marabezi [21], using this methodology converts the masses of glucose, cellobiose and hydroxymethylfurfural into cellulose, through corresponding calibration lines and applying the respective conversion factors: 0.90, 0.5 and 1.7; and the masses of xylose, arabinose and furfural and acetyl groups in hemicelluloses through the same procedure, with a factor of 0.88 for xylose and arabinose, 1.29 for furfural and 0.72 for conversion of acetic acid to acetyl.

The ash content of the wood species was determined by the residue content resulting from the complete burning of the samples, that is, the percentage of inorganic material in the sample, according to the TAPPI T211 om–02 standard [22] (modified). The crushed and dried sample (1g) is added to a porcelain crucible of known mass. This sample was maintained under 525 °C/4 h in a muffle furnace.

The porosity was determined by mercury intrusion in double tests for each wood species, in the Micromeritics Poresizer (model 9320) with 200 MPa pressure, mercury with a surface tension of 0.494 g cm^{-2}, density of 13.533 g/ml, contact angle of advance and retreat of 130° and equilibrium time between low and high pressure of 10 seconds. The properties analyzed were the total pore area (TPA), apparent density (AD) and average pore diameter (APD). First, specimens with nominal dimensions of a 2 cm high prism and 1 cm^2 base were prepared, and then they were placed in a furnace with air circulation at 50 °C for a period of 24 hours for the drying process.

The statistical analysis for the estimation of the physical (TRR, TTR) and mechanical (f_{c0}, f_{t0}, f_{t90}, f_{v0}, f_{s0}, f_{M}, E_{c0}, E_{t0}, E_{M}, f_{H0}, f_{H90}, W, f_{c90}, E_{c90}) properties (Table 2) as a function of wood constituents (He, Cel, IL, SL, Ash, Ext), porosity (Por) and apparent density (ρ_{ap}) for all 10 wood species were obtained with the regression models (Equation (5)) based on analysis of variance (ANOVA) F test ($p < 0.05$), using Minitab® version 16 software.

$$Y = \alpha0 + \alpha1 \cdot He + \alpha2 \cdot Cel + \alpha3 \cdot IL + \alpha4 \cdot SL + \alpha5 \cdot Ash \qquad (5)$$
$$+ \alpha6 \cdot Ext + \alpha7 \cdot Por + \alpha8 \cdot \rho ap$$

where Y denotes the estimated physical and mechanical properties and αi are the coefficients adjusted by the least squares method. The quality of the adjustments was obtained according to the coefficient of determination (R^2). From the ANOVA assumptions of the regression models, p-value below the significance level (p-value < 0.05) implies that the model or its terms are considered significant in the property estimation, and not significant otherwise (p-value>0.05).

Having obtained 12 values of the physical and mechanical properties for each of the ten-wood species, six samples were randomly selected and the average of each group was calculated, resulting in two average values per property and wood species, which were associated with the two sample values of the chemical components and porosity values.

3. Results

Fig 1. shows the results of the three groups for porosity: the highest values (cedroarana, caixeta, cedro and cajueiro), the medim values (cambara, sucupira, cupiúba and cumaru) and the low ones (tatajuba and roxinho); for apparent density: cumaru, tatajuba, roxinho and cupiuba are in the highest values group, sucupira and cambara in the medium one and the lowest are cedroarana, cedro, cajueiro and caixeta.

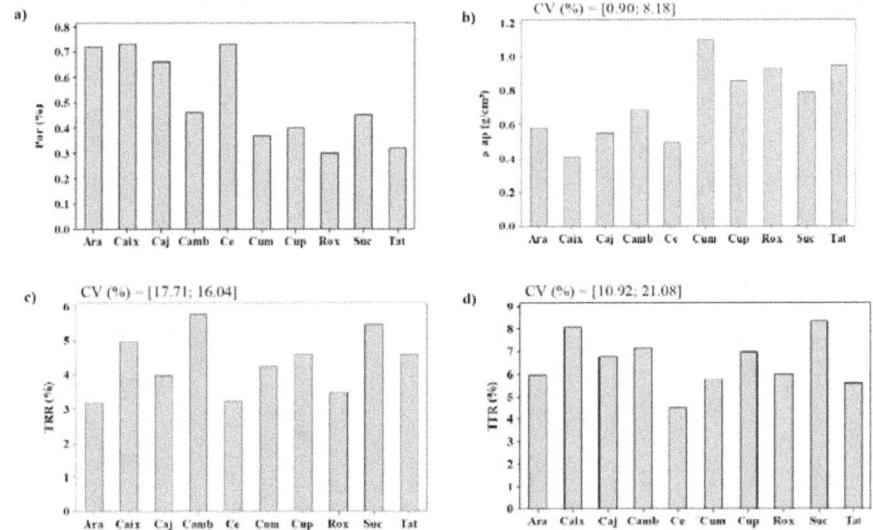

Figure 1. Mean values of the Por, ρap, TRR and TTR properties.

(a) the mean porosity values variations (Por); (b) for the apparent density (ρap); total radial retraction (TRR); and (d) for the total tangential retraction (TTR) as a function of the Ara – Cedroarana; Cax – Caixeta; Caj – Cajueiro; Camb – Cambará; Ce – Cedro; Cum – Cumaru; Cup – Cupiúba; Rox – Roxinho; Suc – Sucupira; Tat – Tatajub. Inserted in these Figure 1a-d upper side the lower and higher values of the coefficient of variation (CV) obtained.

Total radial retraction (TRR can also be divided into three groups: 5% to 6% – caixeta, cambara and sucupira; 4% to 5% – cajueiro, cumaru, cupiuba and tatajuba; 3% to 4% – cedroarana, cedro and roxinho. As with the results for TTR, they can be divided into the following groups: more than 8% – caixeta and sucupira; between 8% and 6% – cajueiro, cambara, cupiuba and roxinho; less than 6% – cedroarana, cedro, cumaru and tatajuba.

Therefore, the ρ_{ap} values obtained were coherent with values of annex E of the Brazilian standard [11], thus the tatajuba and roxinho wood species present the highest density and lowest porosity. The cedro, cedroarana, caixeta and cajueiro wood species present the lower density and higher porosity. Regarding the TRR both in the radial and tangential directions, the cedro and cedroarana wood presented lower values (high stability) than the cambará, sucupira and caixeta wood with lower dimensional stability.

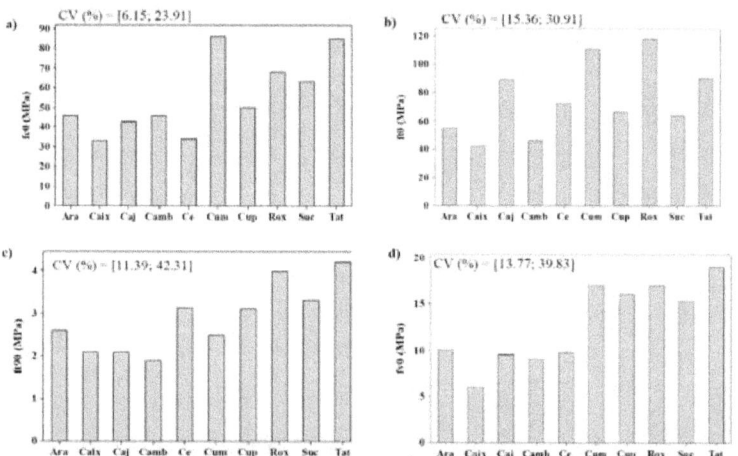

Figure 2. Mean values of fc0, ft0, ft90 and fv0 for wood species.

(a) The variation of values of the strength in compression parallel to the fiber (fc0), (b) tensile strength in parallel to the fiber to the (ft0) and (c) perpendicular (ft90) directions and, (d) shear strength in the direction parallel to the fibers (fv0) as a function for wood species (Ara – Cedroarana; Cax – Caixeta; Caj – Cajueiro; Camb – Cambará; Ce –Cedro; Cum – Cumaru; Cup – Cupiúba; Rox – Roxinho; Suc – Sucupira; Tat – Tatajuba), respectively. Inserted in these Figure 2a-d upper side the lower and higher values of the coefficient of variation (CV) obtained.

14

It can be seen that the f_{c0} values (Fig. 2) can be almost classified into pairs: cumaru and tatajuba (more than 80 MPa), roxinho and sucupira (60 – 70 MPa), cupiuba, cedroarana, cajueiro and cambara (40 – 50 MPa) and caixeta and cedro (30 – 40 MPa). There are four groups for f_{t0}: cumaru and roxinho (100 – 120 MPa), cajueiro and tatajuba (80 – 100 MPa), cedro, cupiuba and sucupira (60 – 80 MPa) and cedroarana, caixeta and cambara (40 – 60 MPa).

Fig. 2 (c) shows that the wood can be classified by: near 4 MPa – roxinho and tatajuba; near 3 MPa – cedroarana, cedro, cupiuba, sucupira and cumaru; near 2 MPa – caixeta, cajueiro and cambara. On the order hand, Fig. 2 (d) seems to have a division in the middle: the different types of wood on the right (cumaru, cupiuba, roxinho, sucupira and tatajuba) have values between 15 and 20 MPa and the woods from on the left (cedroarana, caixeta, cajueiro, cambara and cedro) have values between 5 and 10 MPa.

The coefficient of variation values are similar to the values obtained by other authors [23 – 25]. As can be observed, the fc0, ft0, ft90 and fv0 values obtained were coherent with the values in Annex E of the Brazilian standard [11] for cedro, cumaru, cupiúba, sucupira and tatajuba. The f_{c0} value was significant for the 85 MPa tatajuba wood, as it is very close to 79.5 MPa of annex E of the Brazilian standard [11]. The slight differences are common concerning different batches of the species, according to other studies [26, 27].

Roxinho wood stands out among the three species as it has a higher resistance for the four properties in Fig. 2. Moreover, the denser woods, such as tatajuba and cumaru also showed higher strengths. This is associated to the fact that the density has a strong influence on mechanical properties, as confirmed by other authors [28 – 30].

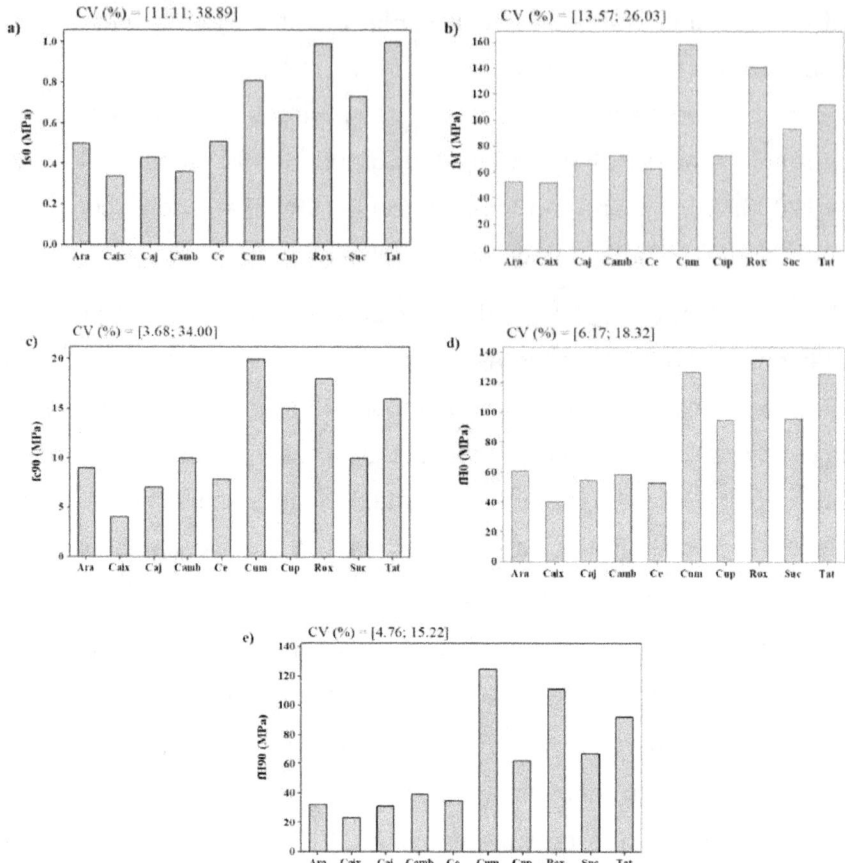

Figure 3. Mean values of fs0, fM, fc90, fH0 and fH90 for wood species.

(a) The variation of the mean values of the strength to cracking (fs0), (b) longitudinal modulus of rupture in static bending (fM), (c) strength in compression perpendicular to the fiber (fc90), (d) hardness of the parallel fiber (fH0) , and (e) hardness of normal fiber (fH90) as a function for wood species (Ara – Cedroarana; Cax – Caixeta; Caj – Cajueiro; Camb – Cambará; Ce – Cedro; Cum – Cumaru; Cup – Cupiúba; Rox – Roxinho; Suc – Sucupira; Tat – Tatajuba), respectively. Inserted in these Figure 3a-e upper side the lower and higher values of the coefficient of variation (CV) obtained.

The variation of the mean values of f_{s0} (Fig. 3) is a property in which the results were diffused, therefore the ascending order for the results is caixeta, cambara, cajueiro, cedroarana, cedro, cupiuba, sucupira, cumaru, roxinho and tatajuba.

On the other hand, Fig 3 (b) shows that the f_M values are mainly between 60 and 80 MPa (cajueiro, cambara, cedro, and cupiuba), but the wood from cedroarana and caixeta have values between 40 and 60 MPa. Furthermore, the f_M values for sucupira, tatajuba, roxinho and cumaru are near 90, 110, 140 and 160 MPa, respectively.

The strength in perpendicular compression to the fiber (f_{c90}), shown in Fig 3 (c), looks like the values for f_{s0} because of the heterogeneity. Therefore, the ascending order for the results is caixeta, cajueiro, cedro, cedroarana, cambara, sucupira, cupiuba, tatajuba, roxinho and cumaru.

For the f_{h0} and f_{h90}, the results are similar as cumaru, roxinho and tatajuba have the highest values, cupiuba and sucupira are in the middle and cedroarana, cedro, caixeta, cajueiro and cambara are the lowest values. Moreover, there is a density relationship that influences these properties (Fig. 3) due to the species with the highest resistance values which were the most dense, similarly obtained in other studies [31, 32].

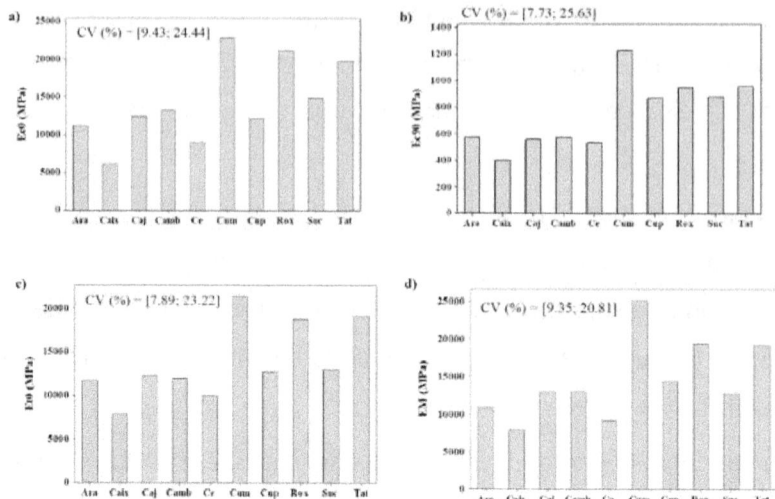

Figure 4. Mean values of Ec0, Ec90, Et0 and EM for wood species.

(a) longitudinal modulus of elasticity in compression parallel to the fiber (Ec0), (b) normal to the fibers (Ec90), (c) in tension parallel to the fiber (Et0), (d) conventional modulus of elasticity in the static bending test (EM) as a function for wood species (Ara – Cedroarana; Cax – Caixeta; Caj – Cajueiro; Camb – Cambará; Ce – Cedro; Cum – Cumaru; Cup – Cupiúba; Rox – Roxinho; Suc – Sucupira; Tat – Tatajuba), respectively. Inserted in these Figure 4a-d upper side the lower and higher values of the coefficient of variation (CV) obtained.

The graph shown in Fig. 4 (a) indicates that this property has dispersed values for the wood, therefore the ascending order for the results is caixeta, cedro, cedroarana, cajueiro, cupiuba, cambara, sucupira, tatajuba, roxinho and cumaru. The Ec90 property of the woods from cedroarana, caixeta, cajueiro, cambara and cedro are between 400 and 600 MPa, the wood from cupiuba, roxinho, subupira and tatajuba are between 800 and 1000 MPa. The only one that is more than 1200 MPa is cumaru.

The longitudinal modulus of elasticity values in Ec0 were coherent with the values according to Annex E of the Brazilian standard [11]. For example, tatajuba has a value of 19.583 MPa according to the standard and, in the present study present a 19.761 MPa was obtained. Such variations are compatible with the values determined by Komariah et al. [33].

The wood from cedroarana, cajueiro, cambara, cupiuba and sucupira have almost the same value for E_{t0}, which is about 12.500 MPa. However, caixeta and cedro have values for this property under 10.000 MPa and roxinho, tatajuba and cumaru are near 20.000 MPa. This was a similar case in Fig 4. (d); the only difference is that the value for EM to cumaru is almost 25.000 MPa.

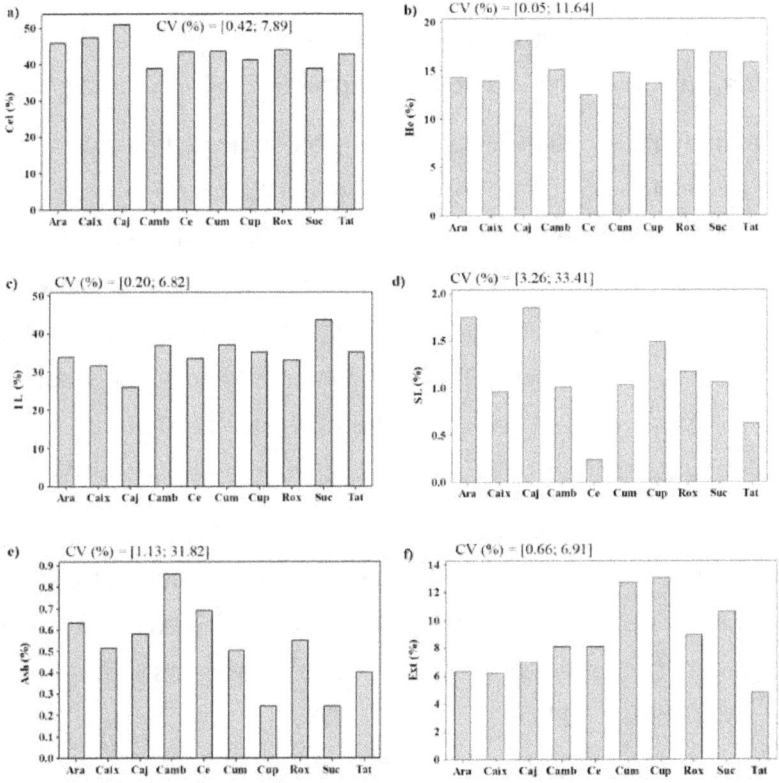

Figure 5. Mean values of wood constituents for all species.

(a) Cellulose (Cel); (b) Hemicellulose (He); (c) Insoluble Lignin (IL); Soluble Lignin (SL); (e) Ashes (Ash); (f) Extracts (Ext); present in the 10 species and the lowest and the highest coefficient of variation (CV) values obtained (upper side inserted).

19

Regarding the chemistry components of these different types of wood (Fig. 5), it can be observed that the ascending orders for the results of each component are: **cellulose** – cambara, sucupira, cupiuba, tatajuba, cedro, cumaru, roxinho, cedroarana, caixeta and cajueiro; **hemicellulose** – cedro, cupiuba, caixeta, cedroarana, cumaru, cambara, tatajuba, sucupira, roxinho and cajueiro; **insoluble lignin** – cajueiro, caixeta, cedroarana, cedro, roxinho, tatajuba, cupiuba, cambara, cumaru and sucupira; **soluble lignin** – cedro, tatajuba, caixeta, cambara, cumaru, sucupira, roxinho, cupiuba, cedroarana and cajueiro; **ash** – cupiuba, sucupira, tatajuba, cumaru, caixeta, roxinho, cajueiro, cedroarana, cedro and cambara; **extractives** – tatajuba, caixeta, cedroarana, cajueiro, cedro, cambara, roxinho, sucupira, cumaru and cupiuba.

The amounts of chemical constituents are in accordance with the values obtained for the 40% to 50% of cellulose [34; 35]; 20% to 30% of hemicellulose [36]; 10% to 36% of total lignin [21]; 2% to 15% of extractives [37]; and 0.1% to 5.0% of ash [38; 39]. The highest concentrations of cellulose and hemicellulose were found in cajueiro wood, which, consequently, presented smaller amounts of total lignin. This fact can be related to the wood aspects, because the cajueiro wood has some similarity with the *Pinus sp.* with regard to density, workability and susceptibility to microorganism attacks [40, 41].

The opposite occurred with sucupira, which had the highest concentrations of total lignin and the lowest concentrations of cellulose and hemicellulose. As lignin promotes stiffness, protection against microorganism attacks and sustainability for fibers and vases, it appears that the wood has high natural durability [42].

Although some studies presented in the literature address the topic of tropical wood extractives, such as those by Santana & Okino [43] allows some comparisons with the results obtained here. It is noteworthy that the types of wood that presented the highest percentages of extractives were Cumaru (approximately 13%) and Cupiúba (approximately 12%), possibly related to the fact that they are the wood species that present the most striking odors among those studied. Immediately after that was Sucupira, with about 10%, which although had no noticeable taste or odor or more intense color, had good durability and resistance to fungi and termites, probably due to the action of the substances contained in the extractives together with the highest concentration of lignin.

Zau et al. [44] show that the chemical, physical and mechanical properties of agglomerated panels produced with cumaru residue, resulted in 18.32% ± 1.24 of organic extractive content. On the other hand, Santana and Okino [43] found 9.7%. Both contents are different from the one found in this study, which shows the need for a greater number of samples analyzed for extraction for the same wood batch, as performed in this study.

Using the same TAPPI T211 standard om-02 [22], Zau et al. [44] found a higher ash content for Cumaru, 1.72% ± 0.15. On the other hand, Silva et al. [45] used the ASTM 1102-56 standard [46] and the ash content value was 0.3%. They point out that the extractive and ash contents may vary due to some factors, such as soil characteristics [44].

Table 3. Results of multivariable regression models applied for the ten wood species.

$Y=\alpha_0+\alpha_1\cdot He+\alpha_2\cdot Cel+\alpha_3\cdot IL+\alpha_4\cdot SL+\alpha_5\cdot Ash+\alpha_6\cdot Ext+\alpha_7\cdot Por+\alpha_8\cdot Pap$	R^2 (%)	F
$TRR = 1.00423 - 0.0598497\cdot He + 0.161199\cdot Cel + 0.211325\cdot IL + 0.219873\cdot SL + 0.54172\cdot Ash + 0.0585386\cdot Ext - 10.3552\cdot Por - 7.89255\cdot Pap$	55.70	NS
$TTR = -13.8972 - 0.17602\cdot He + 0.573686\cdot Cel + 0.489992\cdot IL + 1.0962\cdot SL + 0.434925\cdot Ash + 0.178019\cdot Ext - 19.693\cdot Por - 16.0401\cdot Pap$	86.64	NS
$f_{c0} = -151.933 + 1.24366\cdot He + 1.66847\cdot Cel + 1.42686\cdot IL - 4.5862\cdot SL - 3.30513\cdot Ash - 2.15895\cdot Ext + 25.1747\cdot Por + 106.525\cdot Pap$	99.69	S
$f_{t0} = -120.557 + 7.80428\cdot He + 0.653391\cdot Cel - 3.2484\cdot IL - 25.9725\cdot SL + 4.64536\cdot Ash + 2.29757\cdot Ext + 103.628\cdot Por + 155.129\cdot Pap$	91.94	S
$f_{t90} = 14.3938 + 0.0987351\cdot He - 0.200044\cdot Cel - 0.14628\cdot IL - 0.61742\cdot SL - 2.80478\cdot Ash - 0.154106\cdot Ext + 3.02166\cdot Por + 3.68862\cdot Pap$	70.31	NS
$f_{v0} = 45.0526 + 0.603093\cdot He - 0.946233\cdot Cel - 0.658137\cdot IL - 1.37026\cdot SL - 10.977\cdot Ash - 0.398132\cdot Ext + 21.459\cdot Por + 30.3891\cdot Pap$	96.97	S
$f_{s0} = 0.270621 + 0.0205575\cdot He - 0.00059906\cdot Cel - 0.007045\cdot IL - 0.139357\cdot SL - 0.394968\cdot Ash - 0.0264954\cdot Ext + 0.200725\cdot Por + 1.08872\cdot Pap$	87.99	S
$f_M = -674.814 + 1.34682\cdot He + 12.2643\cdot Cel + 6.38663\cdot IL - 25.2675\cdot SL + 70.657\cdot Ash + 4.15845\cdot Ext - 174.338\cdot Por + 42.228\cdot Pap$	98.99	S
$E_{c0} = -45131.9 + 787.744\cdot He + 431.099\cdot Cel + 205.938\cdot IL - 1500.98\cdot SL + 6479.37\cdot Ash - 103.257\cdot Ext + 2396.52\cdot Por + 26327.9\cdot Pap$	99.28	S
$E_{t0} = -29114.8 + 299.478\cdot He + 365.822\cdot Cel + 72.7035\cdot IL - 1222.82\cdot SL + 3475.32\cdot Ash - 210.424\cdot Ext + 5030.14\cdot Por + 25366.9\cdot Pap$	99.75	S
$E_M = -48886.3 + 83.9516\cdot He + 781.078\cdot Cel + 225.201\cdot IL - 1100.24\cdot SL + 6694.12\cdot Ash + 149.961\cdot Ext - 2691.52\cdot Por + 24469.8\cdot Pap$	99.52	S
$f_{H0} = -105.501 + 0.289965\cdot He + 3.02692\cdot Cel + 1.14436\cdot IL - 8.96208\cdot SL - 18.4164\cdot Ash - 1.04555\cdot Ext - 71.7444\cdot Por + 107.66\cdot Pap$	95.22	NS
$f_{H90} = -472.083 - 1.25237\cdot He + 9.12182\cdot Cel + 4.57334\cdot IL - 17.4205\cdot SL + 30.7517\cdot Ash + 2.20399\cdot Ext - 140.339\cdot Por + 71.3795\cdot Pap$	97.85	NS
$W = -126.674 + 4.894\cdot He - 0.254245\cdot Cel + 1.89813\cdot IL - 5.37704\cdot SL - 20.5034\cdot Ash + 0.855499\cdot Ext + 36.6718\cdot Por - 10.2846\cdot Pap$	99.89	NS
$f_{c90} = 4.48889 - 0.490513\cdot He - 0.0743078\cdot Cel - 0.172129\cdot IL + 0.112121\cdot SL + 2.21648\cdot Ash + 0.0408142\cdot Ext - 3.29173\cdot Por + 23.712\cdot Pap$	97.87	S
$E_{c90} = -1424.82 + 14.22\cdot He + 12.4742\cdot Cel + 6.53891\cdot IL - 72.6691\cdot SL - 107.596\cdot Ash + 10.8421\cdot Ext + 468.118\cdot Por + 1345.81\cdot Pap$	99.89	S

Table 4. Intervals of the values in the independent variables utilized in the regression models.

Properties	Intervals of confiance
He (%)	[12.54; 18.14]
Cel (%)	[38.85; 50.82]
IL (%)	[31.63; 43.63]
SL (%)	[0.24; 1.85]
Ci (%)	[0.24; 0.86]
Ext (%)	[6.17; 13.04]
Por	[0.30; 0.73]
ρ_{ap} (g/cm³)	[0.41; 1.10]

Table 3 shows the regression models and the respective coefficient of determination in the estimation of wood properties, and Table 4 shows the numerical intervals of the independent variables of the obtained adjustments. Moreover, in Table 3, the terms of the models considered significant by ANOVA were duly underlined. The significance of the models was designated as SM, and the models can be significant (S) or not (NS).

It can be observed for the obtained equations, as results, that it presents a good coefficient of determination (R^2 > 90%), except in the three cases, as justified: the first and second cases are the TRR and TTR. As a justification, it should be noted that the variability of these properties is directly related to anatomical aspects of the wood, such as the arrangement of fibers, vessels and medullar rays. In addition, the presence of stalks in the connections between vessel elements may affect the rates of fluid percolation inside the wood, directly interfering with the retraction phenomenon. These aspects are not captured in the statistical analysis because they are not controllable factors, and such effects are not considered. The conclusions of other authors corroborate with this argument [13, 47, 48].

The third case is the resistance of the wood to the normal traction to the fibers. As a justification, it should be noted that this property is strongly influenced by the anatomical characteristics of the wood, in particular, the percentage of amorphous zones at the microfibrillar level and small localized imperfections that, leading to fragile discontinuities, accentuate the character of the typical rupture in this type of request.

It can be observed the according to the NBR ABNT 7190/97, in its article 7.2.3, the safety of structural wooden parts in relation to the limit states must not depend on the normal tensile strength to the fibers, for the reasons stated. Moreover, the mentioned normative document determines that when the normal tensile stresses to the fibers reach significant values in structural situations, devices (generally metallic) must be used that behave as restrictions to the consequences of these stresses. It is clear that these aspects are not captured in the statistical analysis, only with the parameters addressed here. The conclusions of other studies corroborate this argument [13, 47 - 51].

Notice that six properties (E_{c90}, W, E_M, E_{t0}, E_{c0} and f_{c0}) (see Table 3) have a coefficient of determination (R^2) above 99%. which indicates a high level of reliability. The compression strength to parallel fiber (f_{c0}) presents a higher adjustment of R^2 = 99.69%. It is very interesting that this property is the reference, according to the Brazilian standard [11], for the classification of many species not known in the five classes of resistances recommended in this document.

From the analysis of variance, Table 3 shows that the estimation models for the TRR, TTR, f_{t90} , f_{H0}, f_{H90} and W properties, even with R^2 values ranging from 57.70 to 99.89%, were still considered non-significant, and this implies that variations in the dependent variables are not able to explain the variations in the estimated properties, which reveals the small accuracy of the models in the estimation of these properties.

For the ten other properties, the regression models were considered significant by ANOVA, and in all these adjustments, only the apparent density (ρ_{ap}) and porosity (Por) were considered significant, implying that variations in He, Cel, IL, SL, Ci and Ext properties do not explain the variations in f_{c0}, f_{t0}, f_{v0}, f_{s0}, f_M, E_{c0} E_{t0}, E_M, f_{c90} and E_{c90}, which enables us to conclude that the accuracy of the models (R^2) is mostly explained by the apparent density and porosity.

Observing that caixeta, tatajuba and roxinho have the highest apparent density values ($\rho_{ap} > 0.90$ gcm^{-3}), and that 83 % of the native forest types of wood have density values lower than 0.90 gcm^{-3} [52;53], such specimens were excluded for the analysis of f_{c0} via the regression model as f_{c0} is the reference property considered by the Brazilian standard [11] of wood structures. Combining the two types of lignin (IL + SL; total lignin) and excluding the three species with the highest apparent density values, the regression model for f_{c0} estimation resulted in: $f_{c0} = 29 + 1.65 \cdot \text{Cel} - 0.87 \cdot \text{He} - 19.7 \cdot \text{Ci} - 3.500 \cdot \text{Ext} + 24.2 \cdot \text{Por} + 88.7 \cdot \rho_{ap} + 0.01 \cdot (\text{IL} + \text{SL})$, with $R^2 = 97.22\%$.

It should be noted that under these conditions, besides porosity and apparent density, the sum of lignins was also considered significant by ANOVA, indicating that the consideration of specific ranges of apparent density may result in different conclusions from the regression models obtained, and that should be the focus of new research.

It is important to note that the time to obtain the parameters used in the regression models for the wood species is approximately two days, and if such parameters are known, the equations allow the rapid estimation of the various physical and mechanical properties evaluated. This time interval is much lower than that required to promote the complete characterization of the species (destructive testing in the laboratory), usually around three weeks per species, as argued by other authors [32, 54].

4. Discussion

Several authors from different areas of wood engineering observed that the apparent density is a physical property of wood correlated to a strong influence on mechanical properties. For example: as demonstrated by Burdzik and Nkwera [28] in the study on transverse vibration tests to predict stiffness and strength properties of full size *Eucalyptus grandis*; Logsdon et al. [29] in the characterization of cambara's wood; Icimoto et al. [30] researched the influence of lamellar thickness on strength and stiffness of glued laminated timber beams of *Pinus oocarpa*; Rocco Lahr et al. [32] tried to discover the influence of moisture on the strength and stiffness properties of the wood and carryied out the complete characterization of the *Vatairea sp.* wood species.

Cavalheiro et al. [47] tried to find a relationship between apparent density and basic density, as estimators of other physical properties of wood (longitudinal, radial, tangential, volumetric retractions and anisotropy coefficient). To do this, they used coniferous species (*Pinus sp.* and *Pinus oocarpa*) and dicotyledons (Paricá *(Schizolobium amazonicum)*, Jatobá *(Hymenaea sp.)* and Lyptus®) and the linear, exponential, logarithmic and geometric regression models. Lyptus® wood is extracted from renewable forests from planted trees developed with the crossing of species.

After tests using analysis of variance (ANOVA) of the regression models, considered at the confidence level (α) 5%, it was concluded that it was impossible to establish significant relationships to allow the adoption of density as an estimator of other physical properties of wood, due to the fact that they did not find a pattern of behavior both between density and retractions, as well as density and anisotropy coefficient.

In the same research area, Christoforo et al. [48] proposed to correlate the apparent density with physical properties of retraction (radial, longitudinal and volumetric) and coefficient of anisotropy. They used the linear, quadratic,

26

cubic and exponential models, and a species of each of the resistance class established in accordance with the Brazilian standard [11] for the Cedro Doce – Cedrela sp (resistance class C20), Canafístula – Peltophorum dubium. (C30), Angelim Araroba – Vataireopsis araroba (C40), Mandioqueira – Qualea sp (C50) and Angelim Vermelho – Dinizia excelsa (C60). They obtained a value below 0.70 (R²) and indicated that more studies are needed with more species to obtain a value above 0.70 (R²).

Almeida et al. [3] evaluated the correlation between apparent density and compressive strength parallel to the fiber (f_{c0}), establishing an important base for the classification of wood strength as a parameter included in the Brazilian standard [11]. Therefore, wood species such as canafístula (Cassia ferruginea) Angelim araroba (Vataireopsis araroba) and Castle (Gossypiospermum sp.) were utilized, adopting the regression models adopted with no linear, exponential, logarithmic and geometric parameters. However, the regression models produced by Almeida et al. [3] did not provide good estimates for the compressive strength parallel to the fibers, since the best coefficient of determination (R²) was obtained for Canafístula, Angelim araroba and Castle at 48.57%, 14.89% and 52.84%, respectively. Moreover, considering the three species together provided a coefficient of determination of 17.88% (as best estimate).

In this study, in addition to the apparent density property, the porosity parameter and chemical components were used together, which made it possible to obtain more accurate statistical models to correlate with physical and mechanical properties, resulting in the coefficient of determination values (R²) above 95%.

Furthermore, it is known that R^2 is not sufficient to judge the significance of the model. The significance of the model can be assessed by analysis of variance. The progressive inclusion of terms in the models implied proportional increases in the value of the coefficient of determination of all models, implying the significance of each term. However, the analysis of variance by ANOVA was performed, in which 10 properties have significant estimates (fc0, ft0, fv0, fs0, fM, Ec0 Et0, EM, fc90 and Ec90) and only the properties of density and porosity influence significantly in the properties.

Unfortunately, the lack of information in the literature also momentarily impedes the testing of the obtained equations as the values of the chemical properties of the species in the literature differ from the range of values of the results of this study with extrapolation. Therefore, further analysis of the accuracy and validation of the equations may be the subject of future studies.

In addition, due to the limitations inherent in the studies reported in the literature to date, which have a correlation between density, porosity and chemical components, the pioneering nature of this study offers the possibility of continuing research on this topic. In this case, ten wood species were studied, but the number of species can be increased to enhance the reliability of the proposed equations. Moreover, the mechanical properties were observed for the wood presenting moisture content of 12%. This approach can be used to relate the percentages of chemical constituents of the wood with the physical-mechanical properties determined in the condition of moisture equal to or higher than the saturation point of the fibers.

5. Conclusions

From the results of the study it can be concluded that:

1. It is possible to establish reliable equations, from the variables: apparent density, porosity, percentage of cellulose, hemicellulose, insoluble lignin, soluble lignin, extractives and ash of tropical Brazilian wood species, to estimate the set of physical and mechanical properties studied in this research. Therefore, further analysis of the accuracy and validation of equations might be necessary.

2. The accuracy of the estimates provided by the equations was observed by the high values of the coefficient of determination (R^2) indicating that the equations for 10 out of the 16 properties are significant.

3. The parameters to estimate the set of properties studied in this research (of the order of two days per species) can be obtained in time intervals much lower than those necessary to promote the complete characterization of the species through the tests stipulated by the Brazilian standard (usually around three weeks per species).

6. Acknowledgments

This study was financed in part by the Coordination for the Improvement of Higher Education Personnel (CAPES) - Finance Code 001 (Brazil) and National Council for Scientific and Technological (CNPq) - Grant no. #168881/2017-9 (Brazil).

References

[1] Dias F. M. A densidade aparente como estimador de propriedades de resistência e rigidez da madeira. Masters thesis - Interunidades em Ciência e Engenharia de Materiais, Universidade de São Paulo. São Carlos. (2000)

[2] Klock U. et al. Química da Madeira. 3.ed. Curitiba: UFPR. (2005)

[3] Almeida T. H., Ameida D. H., Chistoforo A. L., Chahud E., Branco L. A. M. N., Lahr F. A. R. Density as Estimator of Strength in Compression Parallel to the Grain in Wood. Inter. J. Mater. Eng., 6 (2016) 67 – 71

[4] Celulose Online. http://celuloseonline.com.br/nos-estados-unidos-madeira-e-o-material-mais-importante-para-construir-casas/. Accessed in: 02/17/2017

[5] APA – The Engineered Wood Association. Wood products and other bulding materials used in new residential construction in the United States, with comparison to previous studies. (2012)

[6] Eriksson L. O., Gustavsson L., Hänninen R., Kallio M., Lyhykäinen H., Pingoud K., Pohjola J., Sathre R. Climate change mitigation through increased wood use in the European construction sector–towards an integrated modelling framework. Eur. J. Forest Res. 131 (2012) 131 – 144

[7] Rocco Lahr. F. A., Christoforo A. L., Fiorelli J. Influence of nails size and layout to obtain the reduction coefficient of moment of inertia for timber beams with composite cross section. Eng. Agric. (Online), v. 36, p. 715 – 723, (2016)

[8] Christoforo A. L., Panzera T. H., Araujo V. A., Fiorelli J., Rocco Lahr. F. A. Timber beam repair based on polymer-cementitious blends. Eng. Agric. (Online), v. 37 (2017) 366 – 375

[9] Araújo V. A., Barbosa J. C., Gava M.; Garcia J. N., Souza A. J. D., Savi A. F., Morales, H. A. M., Molina, J. C., Vasconcelos, J. S.; Chistoforo, A. L.; Rocco Lahr. F. A. Classification of Wooden Housing Building Systems. Bioresources (Raleigh, N.C, v. 11 (2016) 7889 – 7901.

[10] Stamade. <http://stamade.com.br/wood-frame>. Accessed in: 05/20/2017.

[11] Associação Brasileira de Normas Técnicas - ABNT. NBR 7190: Projeto de Estruturas de Madeira. Rio de Janeiro. (1997)

[12] Ter Steege H. et al. The discovery of the Amazonian tree flora with an updated checklist of all known tree taxa. Sci. Rep. 6, 29549; doi: 10.1038/srep29549 (Nature) (2016)

[13] Almeida T. H., Almeida D. H., Marcolin L. A., Goncalves D., Christoforo A. L., Rocco Lahr. F. A. Correlation between Dry Density and Volumetric Shrinkage Coefficient of Three Brazilian Tropical Wood Species. Inter. J. Mater. Eng., v. 5 (2015) 50 – 63.

[14] Stolf D. O., Bertolini M. S., Almeida D. H., Silva D. A. L., Panzera T. H., Christóforo, A. L., Rocco Lahr. F. A. Influence of growth ring orientation of some wood species to obtain toughness. Rem-Ver. Esc. Minas, v. 68 (2015) 251– 271.

[15] Rocco Lahr. F. A., Chahud E., Fernandes R. A., Teixeira, R. S. Influência da densidade na dureza paralela e na dureza normal às fibras da madeira. Sci. For. (IPEF), v. 38 (2010) 153 – 158.

[16] Dias F. M.; Lahr F. A. R. Estimativa de propriedades de resistência e rigidez da madeira através da densidade aparente. Sci. For., n. 65, p.102-113, (2004).

[17] Fujimoto T., Matsumoto K., Kurata Y., Tsuchikawa A. Rapid and nondestructive evaluations of wood mechanical properties by near infrared spectroscopy. Japan, (2008).

[18] Silva, D. A. L et al . Influence of wood moisture content on the modulus of elasticity in compression parallel to the grain. Mat. Res., São Carlos , v. 15, n. 2, p. 300-304, Apr. 2012. Available from <http://www.scielo.br/scielo.php?script=sci_arttext&pid=S1516-14392012000200020&lng=en&nrm=iso>. access on 27 Aug. 2018. Epub Mar 06, 2012. http://dx.doi.org/10.1590/S1516-14392012005000025.

[19] TAPPI - Technical Association of the Pulp and Paper Industry. Standart T264 CM-97. Solvent extractives of wood and pulp. (1997)

[20] National Renewable Energy Laboratory – NREL. Determination of Structural Carbohydrates and Lignin in Biomass. Laboratory Analytical Procedure (LAP). (2012)

[21] Marabezi K. Estudo sistemático das reações envolvidas na determinação dos teores de lignina e holocelulose em amostras de bagaço e palha de cana-de-açúcar. 142 f. Masters thesis - Instituto de Química de São Carlos, Universidade de São Paulo, São Carlos. (2009)

[22] TAPPI - Technical Association of the Pulp and Paper Industry. Standart T211 om-02. Ash in wood, pulp, paper and paperboard: combustion at 525°C. (2002)

[23] Arwade S., Winans R.; Clouston P. Variability of the compressive strength of parallel strand lumber. J. Eng. Mech., v. 136, n. 4 (2010) 405 – 412.

[24] Ravenshorst G., Van de Kuilen J. Comparison of methods of strength classification of (tropical) hardwood timber. 11th World Conference on Timber Engineering 2010, WCTE 2010, v. 2 (2010) 1004 – 1012,

[25] Edwin P., Ozarska B. Bending properties of hardwood timbers from secondary forest in Papua New Guinea. J. Trop. For. Sc, v. 27, n 4 (2015) 456 – 461.

[26] Rocco Lahr. F. A., Christoforo A. L., Silva C. E. G., Andrade Junior J. R., Pinheiro R. V. Avaliação de propriedades físicas e mecânicas de madeiras de Jatobá (Hymenaea stilbocarpa) com diferentes teores de umidade e extraídas de regiões distintas. Rev. Arvore, v. 40 (2016) 147 – 154,

[27] Christoforo A. L., Arroyo F. N., Silva D. A. P. L., Panzera T. H., Rocco Lahr. F. A. Physico-mechanical characterization of the Anadenanthera colubrina wood specie. Eng. Agric. (Online), v. 37 (2017) 376 – 384

[28] Burdzik W. M. G., Nkwera P. D. Transverse vibration tests for prediction of stiffness and strength properties of full size Eucalyptus grandis. Forest Prod. J., v. 52, n. 6 (2002) 63 – 67.

[29] Logsdon N. B., Finger Z., Rosa L. M. Caracterização da madeira de Cambará, Vochysia guianensis Aubl. Rev. Eng. Civil, n. 29 (2007) 57 – 69.

[30] Icimoto F. H., Calil Neto C., Ferro F. S., Macedo L. B., Christoforo A. L., Lahr F. A. R., Calil Jr C. Influence of lamellar thickness on strength and stiffness of glued laminated timber beams of Pinus oocarpa. Int. J. Mater. Eng., v. 6 (2016) 51 – 55.

[31] Logsdon, N. B., Calil Jr. C. Influência da umidade nas propriedades de resistência e rigidez da madeira. Cad. Eng. Estruturas, n.18 (2002) 77 – 107.

[32] Rocco Lahr. F. A., Aftimus B. H. C., Arroyo F. N., Christoforo A. L., Chahud, E., Branco L. A. M. N. Full Characterization of *Vatairea sp.* Wood Specie. Inter. J. Mater. Eng., v. 6 (2016) 92 – 96,

[33] Komariah R. N., Hadi Y. S., Massijaya M. Y., Suryana J. Physical-mechanical properties of glued laminated timber made from tropical small-diameter logs grown in Indonesia. J. Korean Wood Sc. Technol., v. 43, n. 2 (2015) 156 – 167

[34] Miller, R.B. Wood handbook: wood as an engineering material. Madison, Wisconsin, Forest Products Laboratory – FPL (1999) p. 463.

[35] Sjostrom S. Wood chemistry: fundamentals and applications. London: Academic Press, Inc., (1981).

[36] Fengel D., Wegener G. Wood: chemistry, ultrastructure, reactions. Berlin: W. de Gruyter, (1984).

[37] Silva M. R. Efeito do tratamento térmico nas propriedades químicas, físicas e mecânicas em elementos estruturais de Eucalipto citriodora e Pinus Taeda. Doctoral dissertation. Programa de Pós-Graduação Interunidades em Ciências e Engenharia de Materiais e Área de concentração em Desenvolvimento Caracterização e Aplicação de Materiais. Escola de Engenharia de São Carlos; Instituto de Física de São Carlos; Instituto de Química de São Carlos, da Universidade de São Paulo, (2012).

[38] Mitchual S. J., Mensah K. F., Darkwa N. A. Evaluation of Fuel Properties of Six Tropical Hardwood Timber Species for Briquettes. J. Sustainable Bioenergy Systems, v. 4 (2014) 1 – 9.

[39] Nisgoski S., Magalhães W. L. E., Batista F. R. R. Anatomical and energy characteristics of charcoal made from five species. Acta Amazon. (2014).

[40] Freitas J. F. De Souza A. M., De Granco L. A. M. N., Chahud, E., Christoforo, A. L., LAHR F. A. R. Production of Structural OSB with Cajueiro (Anacardium sp.) and Amescla (Trattinikia sp.) - A Preliminary Study. Inter. J. Mater. Eng., v. 7(1) (2017) 17 – 20.

[41] REMADE. < http://www.remade.com.br/madeiras-exoticas/148/madeiras-brasileiras-e-exoticas/caju>. Accessed in: 09/14/2017.

[42] REMADE. < http://www.remade.com.br/madeiras-exoticas/445/madeiras-brasileiras-e-exoticas/sucupira >. Accessed in: 09/14/2017.

[43] Santana M. A. E., Okino E. Y. A. Chemical composition of 36 Brazilian Amazon forest wood species. Holzforschung, Berlin. (2007).

[44] Zau M. D. L., Vasconcelos R. P. De Giacon V. M., Lahr, F. A. R. Avaliação das propriedades química, física e mecânica de painéis aglomerados produzidos com resíduo de madeira da Amazônia - Cumaru (Dipteryx Odorata) e resina poliuretana à base de óleo de mamona. Polímeros, São Carlos, v. 24, n. 6 (2014) 726 – 732.

[45] Silva G. A. C., Varejão M. J. C., Nascimento C. C. Estudos tecnológicos de alternativas de uso de resíduos madeireiros. Anais do 61ª Reunião Anual da SBPC, Manaus, AM. (2009) p. 5356.

[46] ASTM International. Standard D 1102-56. Standard Test Method for Ash in Wood. 1978.

[47] Cavalheiro R. S., Almeida D. H.; Almeida T. H., Christoforo A. L., Lahr F. A. R. Density as Estimator of Shrinkage for Some Brazilian Wood Species. Inter. J. Mater. Eng., v. 6(3) 2016) 107–112.

[48] Christoforo A. L., Almeida T. H., Almeida D. H., Santos J. C., Panzera T. H., Lahr F. A. R. Shrinkage for some wood species estimated by density. Inter. J. Mater. Eng., v. 6 (2016) 23 – 27.

[49] Ardalany M., Deam, B., Fragiácomo M., Crews, K. I. Tension perpendicular to grain strength of wood, Laminated Veneer Lumber (LVL), and Cross-Banded LVL (LVL-C). Incorporating Sustainable Practice in Mechanics of Structures and Materials - Proceedings of the 21st Australian Conference on the Mechanics of Structures and Materials, (2011) 891 – 896.

[50] Kretschmann, D. E., Mascia, N. T. Testing of small clear tension perpendicular to grain samples of Sugar Maple for radial, tangential and 45 degree loading orientations. Source: WCTE 2016 – World Conference on Timber Engineering, (2016).

[51] Widmann R., Fernandez –Cabo J. L., Steiger, R. S. Mechanical properties of thermally modified beech timber for structural purposes. Eur. J. Wood-Wood Prod., v. 70, n. 6 (2012) 775 – 784.

[52] INSTITUTO DE PESQUISAS TECNOLÓGICAS. Métodos de ensaios adotados no I.P.T. para o estudo de madeiras nacionais. Tabelas de resultados obtidos para madeiras nacionais. Nomenclatura das madeiras nacionais. 2ª ed. são Paulo, IPT, 1956. 62 p. (Boletim Técnico, 31).

[53] INSTITUTO BRASILEIRO DO MEIO AMBIENTE E DOS RECURSOS NATURAIS RENOVÁVEIS – IBAMA. Madeiras da Amazônia: características e utilização - Amazônia Oriental. Brasília: CNPq, 1997b. vol. 3. 141p

[54] Christoforo A. L., Arroyo F. N., Silva, D. A. L., Panzera T. H., Rocco Lahr. F. A. Full Characterization of Calycophyllum Multiflorum Wood Specie. Eng. Agri., v. 37 (2017) 637 – 643.

Publisher: Eliva Press SRL

Email: info@elivapress.com

www.ingramcontent.com/pod-product-compliance
Lightning Source LLC
Chambersburg PA
CBHW051300170526
45165CB00004B/1792